THE SILENCER COOKBOOK

.22 RIMFIRE SILENCERS

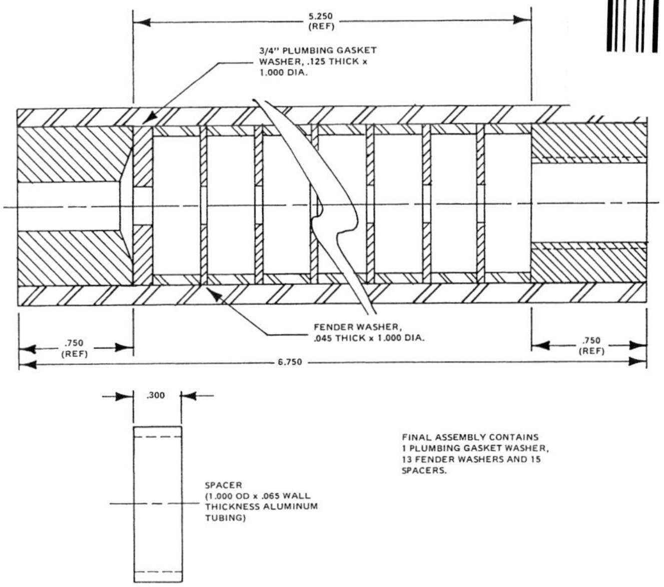

Desert Publications
El Dorado, AR 71730

The Silencer Cookbook

.22 Rimfire Silencers

215 S. Washington
El Dorado, AR 71730
870-862-2077
info@deltapress.com

ISBN 0-87947-057-7
15 14 13 12 11 10 9
Printed in U. S. A.

Desert Publications is a division of
The DELTA GROUP, Ltd.
Direct all inquiries & orders to the above address.

Neither the author nor the publisher assumes any responsibility for the use or misuse of the information contained in this book.

Warning!!

The Publisher (Desert Publications) produces this book for informational purposes and under no circumstances advises encourages or approve of use of this material in any manner.

Table of Contents

WARNING

The manufacture and ownership of sound suppressors in strictly regulated by the federal government. Some state laws and local ordinances flat out forbid them, even though the potential manufacturer or owner may have fulfilled all the requirements for legitimate ownership under federal law. The information contained herein is for educational and reference purposes only

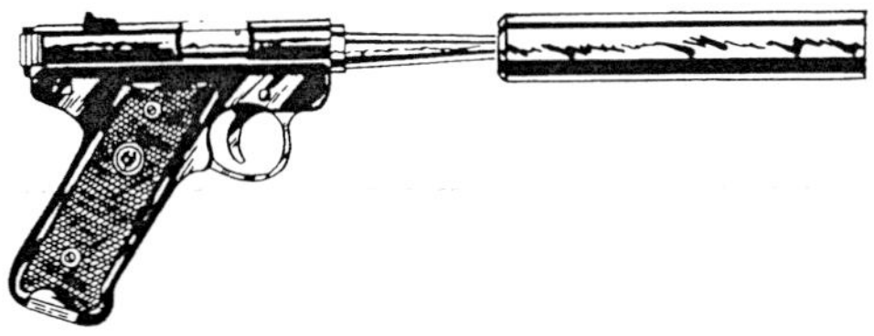

Historical Background

Firearm sound moderators and suppressors, commonly known as "silencers" to most people, have a certain mystique about them that can't be described, only felt. To the person who has a rabid hatred of guns, the silencer's mystique is similar to that of a coiled viper: fascinating in a deadly, evil sort of way. To the average shooter who knows what they are but has never used one, they conjure up visions of a foggy night on the waterfront in Istanbul, with a Soviet spy meeting his end as a CIA agent's silenced pistol meets out whispering death. To a person who is mechanically inclined but who may or may not be a shooter, silencers are fascinating from the standpoint of design and construction and how and why they do or don't work.

Although attempts had been made to silence firearms for several centuries, successful silencers are a product of the Twentieth Century. The large amount of powder residue resulting from firing black powder loads would have quickly fouled any silencer made 100 years or so ago, no matter how functional its design or how high a quality its workmanship. With the advent of modern smokeless powders in the 1890's, the way was open for someone to develop the first truly effective silencer.

The man who can be credited with this was Hiram P. Maxim, son of the man who invented the overhead sprinkler system of fire extinguisher and several types of machine guns, Sir Hiram S. Maxim. Hiram P. patented his first silencer in March of 1908. It was a rather crude, ineffective device, but it was a step in the right direction. Two other Maxim designs followed in 1909 and 1910, with the latter probably being the most popular silencer ever marketed. Numerous other inventors patented silencer designs between 1908 and 1920, but the Maxims are the only ones anyone remembers or hears of today.

Although they probably had a strong appeal to persons with a mechanical inclination, silencers actually had no mystique back in the days when they were legal. If you wanted to shoot rats in the chicken house without disturbing the neighbors, you simply went down to the local hardware store and bought yourself a silencer. If they didn't have any in stock, you could order one through the mail. If you wanted to do a little target practice in the basement without disturbing a napping child upstairs, you simply screwed a silencer onto the end of the barrel of your rifle or handgun and popped off a few rounds at your leisure.

Early Maxim catalogs played up this leisure time use of silencers for everything from informal trap shooting with smoothbore rifles on the front lawn to hunting. Letters from customers extolling the Maxim silencer were often included. Hunters liked it because if they missed their quarry with the first shot from a silenced weapon, their location wasn't revealed by the report of a shot. They could easily get a second or even a third chance, which certainly wasn't so with a non-silenced rifle, the location of which could be easily pinpointed.

One might think that Hiram Maxim manufactured firearm silencers right up until the time they were outlawed, but this isn't so. Since Maxim was interested in silencing noise pollution of all types, he had diversified into producing industrial silencers and diesel engine mufflers as well. These evidently proved far more profitable than firearm silencers, for the latter were dropped from the Maxim line in 1925.

During the heyday of silencers from 1910-1934, Maxim only had two real competitors for the American silencer market, Parker-Hale of England and the old New England arms firm of Hopkins & Allen. The latter's silencer was based on a design patented by Charles H. Kenney in 1910.

In a round about way, the one person most responsible for the gun control laws and high crime rate we have today was a Minnesota Congressman named Andrew John Volstead. He introduced the National Prohibition Act of 1919, which was enacted as the 18th Amendment to the Constitution. That act forbade the manufacture, sale or consumption of intoxicating liquors in the United States. What it actually accomplished was to enable mobsters like Al Capone and Bugs Moran to amass tremendous personal fortunes (Capone's income from all illicit sources during Prohibition has been estimated at as high as $1,000,000 a day) and, of course, power and organization. There was a lot of money to be made from illegal booze, whether it was making the stuff in a backwoods still, running it into the country by boat, or selling it to the customer in speakeasies.

Naturally, any mob that was big enough to have an actual territory also wanted the territory next door, and they weren't about to walk into the neighborhood real estate office to make a cash offer for it. "Offers" were usually threats which were carried out if the rival gang's leader didn't take the hint that he should retire "for health reasons". In most instances, once the threats were carried out the victims usually didn't have to worry about their health or anything else – ever again. Such terminal results were apt to inspire retaliation in kind. Florists in Chicago, New York and other cities with large mob concentrations did a booming business in floral arrangements for funerals.

Many, although certainly not all, of the mob "hits" were made with Thompson submachine guns. It didn't take very many such hits before people began to decry this awesome gangland weapon which had, of course, actually been developed for military use. The mob bloodbath lasted for only a few short years, but gangland killings still occur frequently enough today to remind us that organized crime is everywhere.

By the early 1930's, it was obvious to even the most diehard temperance advocates that Prohibition was an utter, total failure – period. Attempts to create a "dry" America had turned it into a drinking nation. Typical of the more perverse side of human nature, people who had never taken a drink before suddenly developed a taste for alcohol once it became illegal. The 21st Amendment to the Constitution was quickly passed and Prohibition was repealed in 1933.

Following repeal, the politicians and bureaucrats found themselves faced with new problems. The Great Depression was in full swing, with millions out of work and banks foreclosing on homeowners, businesses and farms. With the profits from illicit booze gone, the mobs had turned to gambling, prostitution, loan sharking and even legitimate business as sources of income. A new breed of gangster was capturing the headlines – the bank robbers. To many Americans who had been through foreclosures and who were sick to death of banks, people like John Dillinger, Bonnie and Clyde, Pretty Boy Floyd, Baby Face Nelson and Machine Gun Kelly almost seemed like avenging angels in some distorted form. They hit the banks where it really hurt – in the vault. Unlike the mobsters of the previous decade who often had large estates and offices in the downtown financial district, the small bank robber gangs had no home base and were highly mobile. They might hit a bank in Louisiana on Monday, another in New Mexico or Arizona on Friday and yet another in Nebraska the following week. They did, however, have one thing in common with the highly organized mobs of the Capone era – they used guns. Among their favorite weapons were Thompsons and sawed off shotguns.

In an attempt to stop the crime wave that was as laughable as the attempt to prohibit alcohol, Congress passed the National Firearms Act of 1934. This law made it illegal to manufacture, own or transfer certain types of weapons and "other devices" without prior approval of Federal authorities. To further discourage traffic in such items, a $200 transfer tax was imposed, quite a sum in those days. People who already owned the prohibited items prior to the act's passage were given a certain length of time to register them without paying the transfer tax. Among the newly regulated goodies were automatic weapons of all types, sawed off rifles and shotguns, and silencers.

As was to be expected, only honest citizens complied with the law. The gangsters could have cared less about breaking it. The law had little effect in stopping the bank robbers, since most of them were already dead or in prison when it was passed. The last of the breed, Ma Barker, was killed in a shootout with the FBI in January 1935, only a few months later. NFA '34 did, however, put a crimp in the ownership of certain exotic weapons by the average citizen, to whom a $200 transfer tax might well represent two months' pay at that time.

As a final irony, remember that this act which was supposed to help control crime was passed in an era when it was still safe to walk the city streets alone at night and when most people could leave their houses unlocked even when they were away. A lone citizen was probably safer on the waterfront then than he or she would be in Central Park today.

With the open market for silencers gone, it's no wonder that the few silencer manufacturers either went out of business or else changed their product lines. When the United States entered World War II and there was suddenly a demand for silencers for covert operations, new manufacturers had to be found. The Maxim Silencer Company made up one silenced rifle prototype at the O. S. S's request, but it was never adopted.

Although the O.S.S. used several different types of silenced weapons during the war, the only ones that need concern us here are the .22 rimfire models. The most commonly encountered was the High Standard HD Military. The High Standard was basically just a standard .22 autoloader of good quality fitted with a thin, perforated barrel. This barrel was then wrapped in bronze screen. A silencer tube with bronze washers forward of the barrel was then fitted over it. This simple arrangement was highly effective as was so aptly demonstrated to President Franklin Delano Roosevelt in an early wartime interview.

O.S.S. Chief Gen. William "Wild Bill" Donovan took a sample of the new silenced High Standard to the White House to show the President. FDR was dictating when Donovan entered the room, so he set a sandbag in one corner of the room and then proceeded to empty the pistol's magazine into it. FDR had heard nothing, and was astonished when Donovan handed him the empty, still-smoking pistol.

There is, of course, no written public history of the use of the silenced High Standard during World War II, but it is known that the weapon saw service in Europe and the China-Burma-India Theater. Just how many American agents survived the war thanks to their silent partner isn't known, but one can well guess that it was more than a mere handful.

A slightly different version of the High Standard was made up for the C.I.A. following the war and so far as is known, is still in use today by some operatives. Famed U2 pilot Frances Gary Powers was carrying a silenced High Standard in his escape kit when he was shot down over the Soviet Union during the Eisenhower Administration.

The 0.S.S. also used the Colt Woodsman, at least one example of which was fitted with an eccentric silencer that looks similar to the Maxim Model 1910. This weapon is rather unusual in that it has a barrel extension projecting beyond the silencer. Little else is known about the O.S.S.'s use of the Colt.

Surprisingly, silencer design and production isn't a dead art today, but is actually thriving in a small way.

There are two reasons for this.

The first is that silenced .22 pistols have become a favorite with mob hit men. One might think that professional killers would go for the stopping power Of the .45 ACP or .44 Magnum rounds. However, their actions are a whole different ball game than defensive shooting. The hit man always strikes from close range and usually has the victim under his direct control, so it's no great task to pump a half dozen .22 slugs into the back of the victim's head. A silenced .22 is a lot quieter than a silenced .45, which adds up to less attention being drawn to the scene of the hit. This in turn increases the killer's chances of escaping detection. The "professional" demand for silenced .22 pistols is sufficient to warrant several underground factories turning them out or even offering "hit kits" consisting of an attache case containing a .22 pistol which can be fired while still inside the case. While just about any .22 autoloader can be used for this purpose, the Ruger ST seems to be a favorite, and not just because it's one of the lowest priced .22's on the market. It's also one of the most reliable.

The second reason for the current state of silencer design and manufacture is that there is a growing number Of private citizens who are interested in acquiring a legal silencer for their personal use. The $200 transfer tax amounts to less than a week's pay for many people today, so isn't the deterent to legal ownership that it once was.

Many prospective silencer owners admittedly just want a neat new toy that no one else on their block has, but a number also have a legitimate use for one. The old saying about wiping out the rats in the henhouse without affecting egg production isn't as silly as it sounds. Farmers and ranchers often have to worry about keeping the critter population on their land under control without scaring the livestock. Silencers are also useful on indoor ranges, where the sound of a gun's report is understandably louder than outdoors. Even when the shooter is wearing proper ear protection, the use of a silencer for indoor shooting can well be justified.

When reviewing prospective silencer purchasers, one also needs to consider a special group – the survivalists. If any of their scenarios for the collapse of our society should come to pass, a silencer for poaching or taking out enemy patrols might literally make the difference between life and death. People who are unfamiliar with survivalists tend to look on them as 'lust another nut group", but there is really no way to classify survivalists as a group except that they all wish to survive whatever calamity might change civilization and the social structure as we know them. Some survivalists are opposed to guns and don't own any, but most own a number. These often include military style rifles and handguns for defense, general type rifles and shotguns for hunting and specialized 'lust in case" weapons such as the Charter Arms AR-7 shown herein. The AR-7 was the first true survival gun to be marketed for the general public and it has been popularly accepted. Its most unique feature is the fact that it can be disassembled and the barrel and receiver stored in the hollow buttstock. This makes the AR-7 ideal to just leave in the trunk of a car or to carry in a boat or canoe.

The silencers that appear on the following pages were all made from commonly available materials, most of which can be purchased at the local hardware store. None of them require elaborate machining or other processes which are beyond the capability of the average person who is limited to a few basic tools. They were assembled and tested by a licensed Class II manufacturer, so are known to effectively reduce the sound of a .22's report.

Naturally, anyone who wishes to duplicate these silencers should fill out the necessary paperwork and submit it for BATF approval prior to attempting construction. Even if the silencer is to be used for purely mundane purposes, it is still illegal and the owner is subject to prosecution unless proper authorization from the BATF was given *before* the silencer was made.

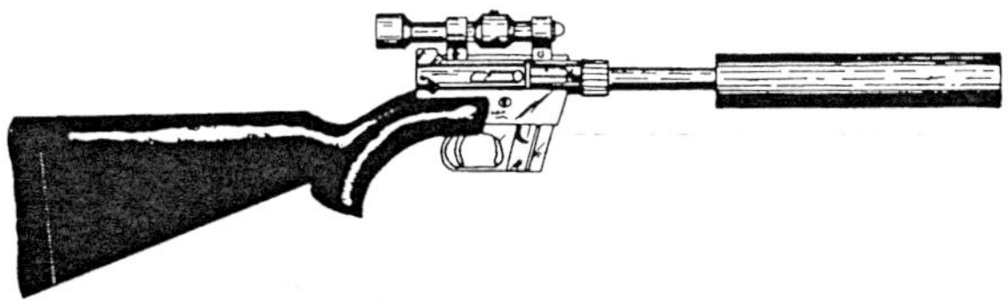

Construction Details

This is a how-to book. You are given several choices for the construction of a .22 rimfire caliber suppressor. Since the outer tube of the suppressor is the registered part, the "insides" can be changed or altered at will. With this premise in mind we have fashioned several suppressor designs, all fitting the same basic tube.

The only option concerning the tube is whether you use threaded end plugs or use a simplified method of fastening the end plugs with set screws. While the latter may be easier to construct, the threaded end plugs with flanges will insure better alignment with the barrel. If you don't have access to a lathe you will have to depend on an outside machine shop to produce the parts for you. It is wise to "shop around" for an economical machinist. If you live in an area where machine shops service the aerospace industry, you will find their rates out of sight when compared to the hobby machinist with tools in his garage.

The drawings in this book were produced for you to reproduce on a copy machine. They were, also, made to appear as innocuous as possible, not for illegal purposes, but to avoid alarming the uninformed. Having BATF approval will not necessarily guarantee that everyone shares your interest in the exotics of weaponry. The author once had a wood working shop refuse to make a small run (100) of wooden stocks for the Uzi semi-auto carbine. This well-meaning nerd was afraid that these stocks would "fall into bad hands"!

A good cover story for a shop where you're not known is that you are an inventor and these parts are for a prototype you're building. Everyone expects inventors to be secretive about their work. Also, inventors are known for being poor, so you may be charged a minimum rate for the work done.

What we're advising you to do is not intended to help circumvent the law, only help you avoid pitfalls when doing things legally.

PLAIN BAFFLE SUPPRESSORS

The most easily obtained baffles are made from fender washers. The ones used here were originally intended to fit over a 3/16" dia. screw. Their OD was a few thousandths over 1.000. Since the tubing we used had a wall thickness 0f .120, the inside diameter was 1.009. The fender washers fit perfectly, although the center hole had to be enlarged to at least .250.

The alignment of any type suppressor baffle is affected by how precision its outside diameter is, compared to the inside diameter of the tube. The looser the fit the larger the clearance hole for the bullet will have to be.

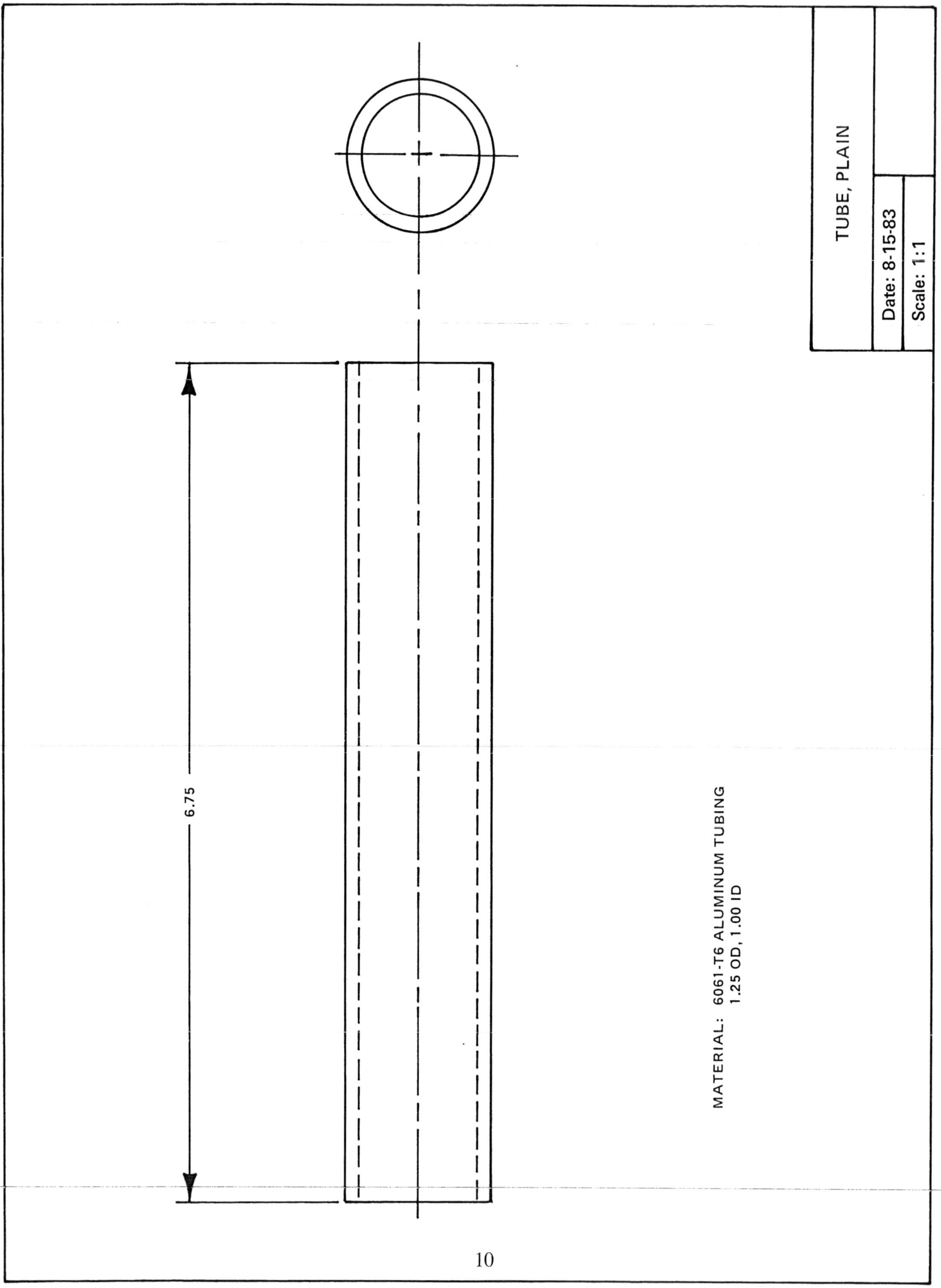
TUBE, PLAIN
Date: 8-15-83
Scale: 1:1
6.75
MATERIAL: 6061-T6 ALUMINUM TUBING
1.25 OD, 1.00 ID

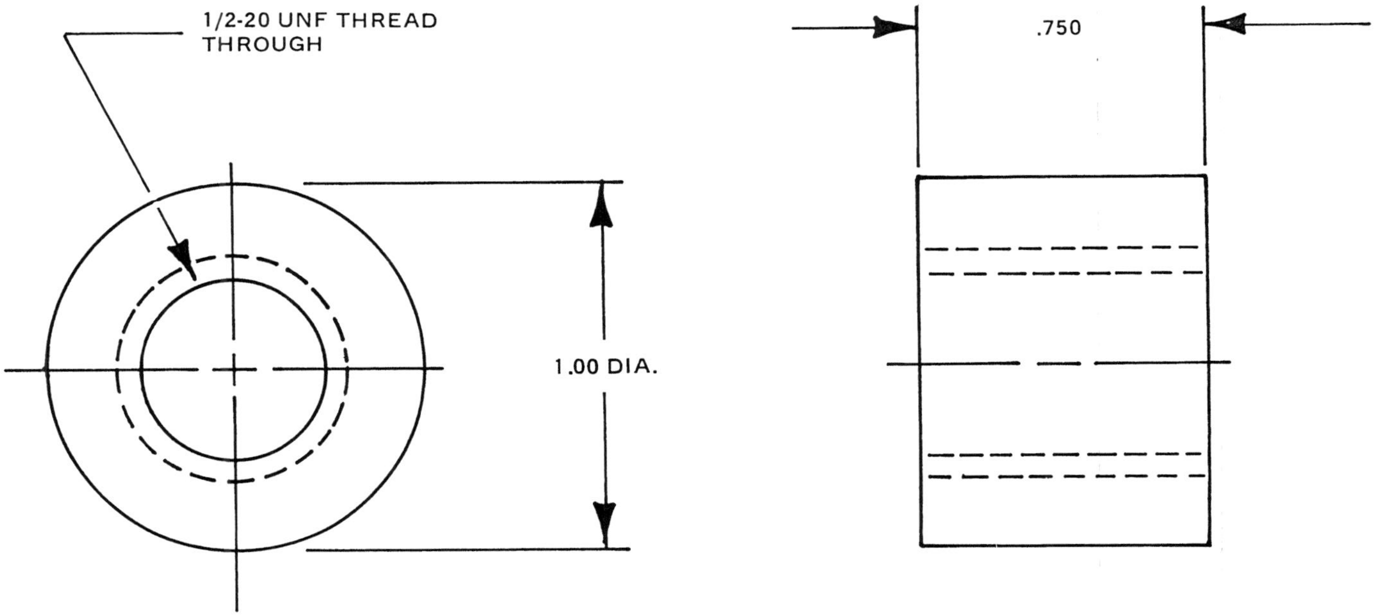

MATERIAL: 2024-T3 ALUMINUM ROUND STOCK.

BREAK ALL SHARP EDGES.

END PLUG, REAR	
Date: 8-15-83	
Scale: 2:1	

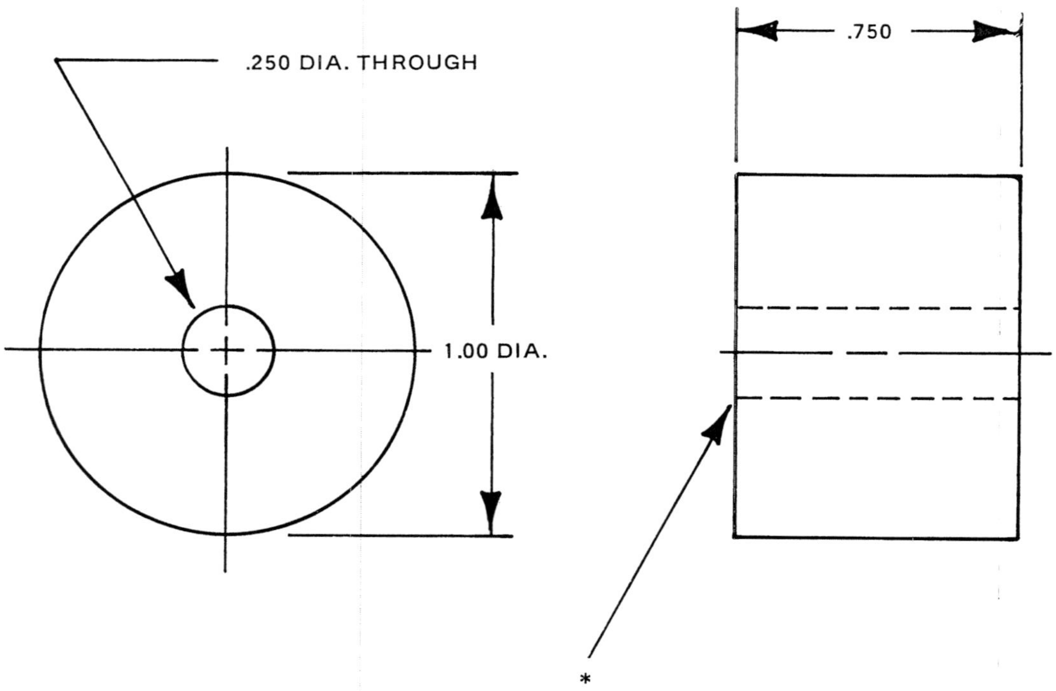

* SEE PAGE 14 FOR OPTIONAL SHAPES FOR THIS SURFACE.

MATERIAL: 2024-T3 ALUMINUM ROUND STOCK.

BREAK ALL SHARP EDGES.

END PLUG, FRONT	
Date: 8-16-83	
Scale: 2:1	

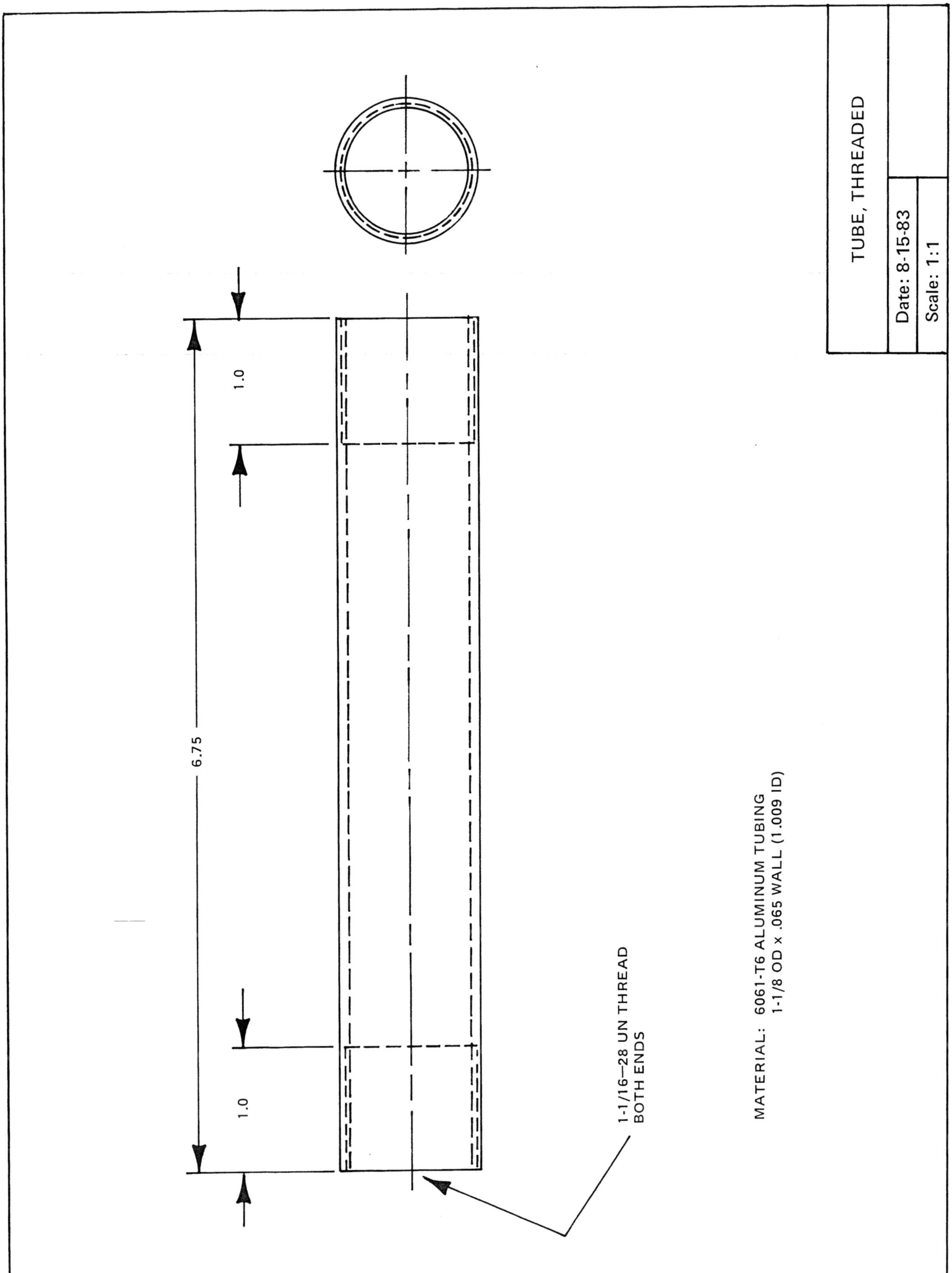
TUBE, THREADED
Date: 8-15-83
Scale: 1:1
1.0
6.75
1.0
1-1/16—28 UN THREAD
BOTH ENDS
MATERIAL: 6061-T6 ALUMINUM TUBING
1-1/8 OD x .065 WALL (1.009 ID)

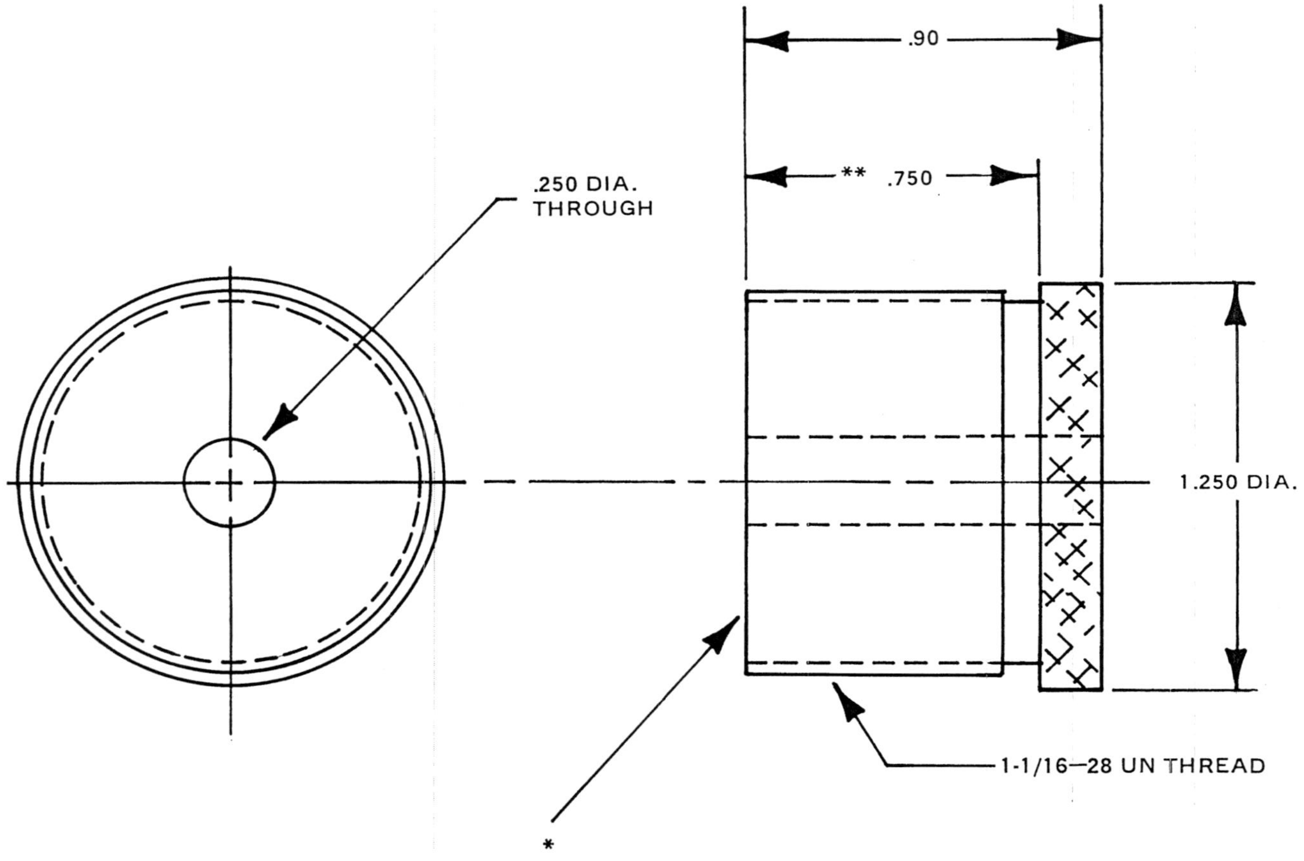

MATERIAL: 2024-T3 ALUMINUM ROUND STOCK.

* SEE PAGE 14 FOR OPTIONAL SHAPES FOR THIS SURFACE.

** THIS DIMENSION CAN VARY TO FIT PARTICULAR REQUIREMENTS

BREAK ALL SHARP EDGES.

END PLUG, FRONT	
Date: 8-16-83	
Scale: 2:1	

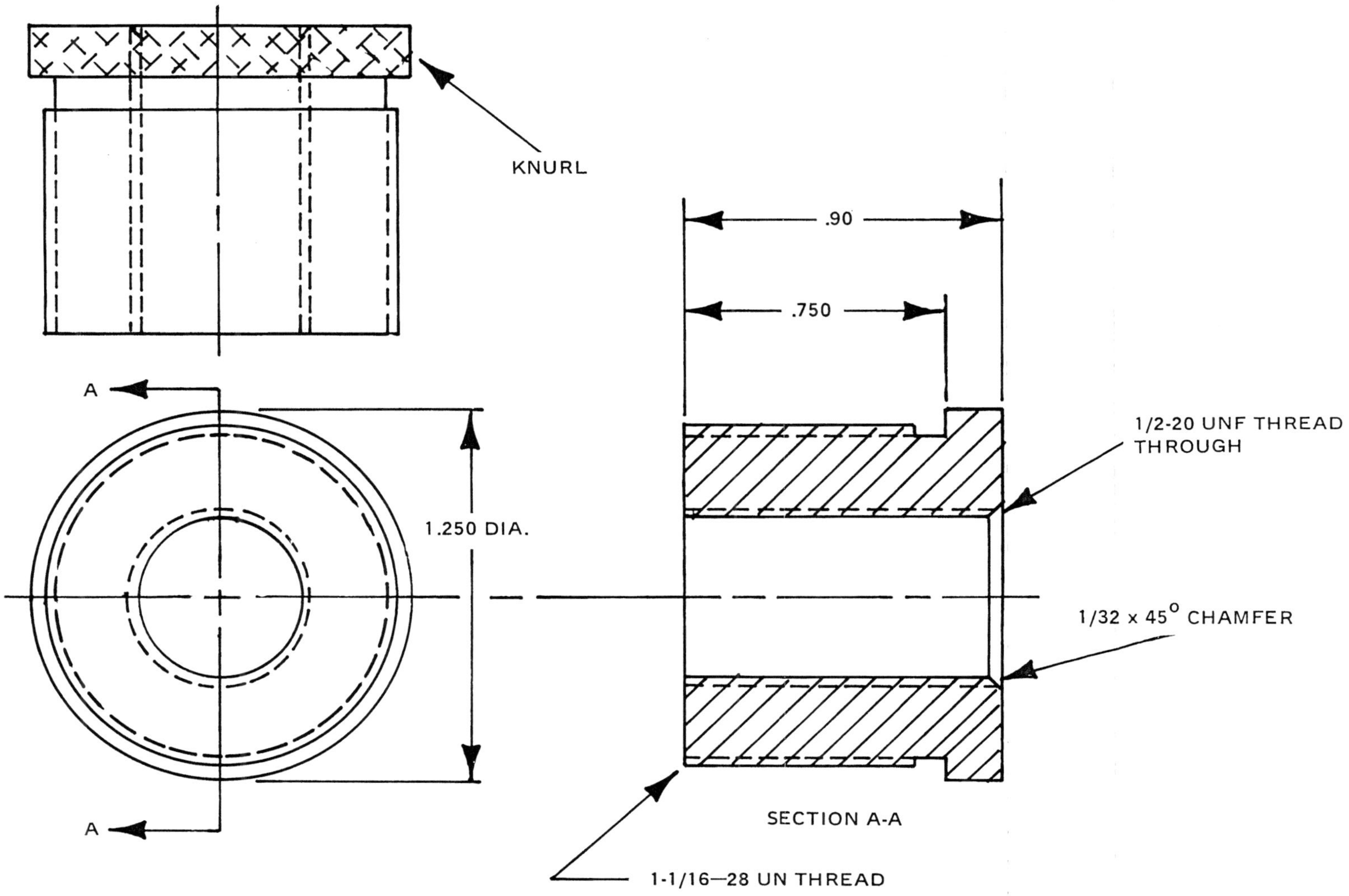

MATERIAL: 2024-T3 ALUMINUM ROUND STOCK.

BREAK ALL SHARP EDGES.

REAR END PLUG	
Date: 8-16-83	
Scale: 2:1	

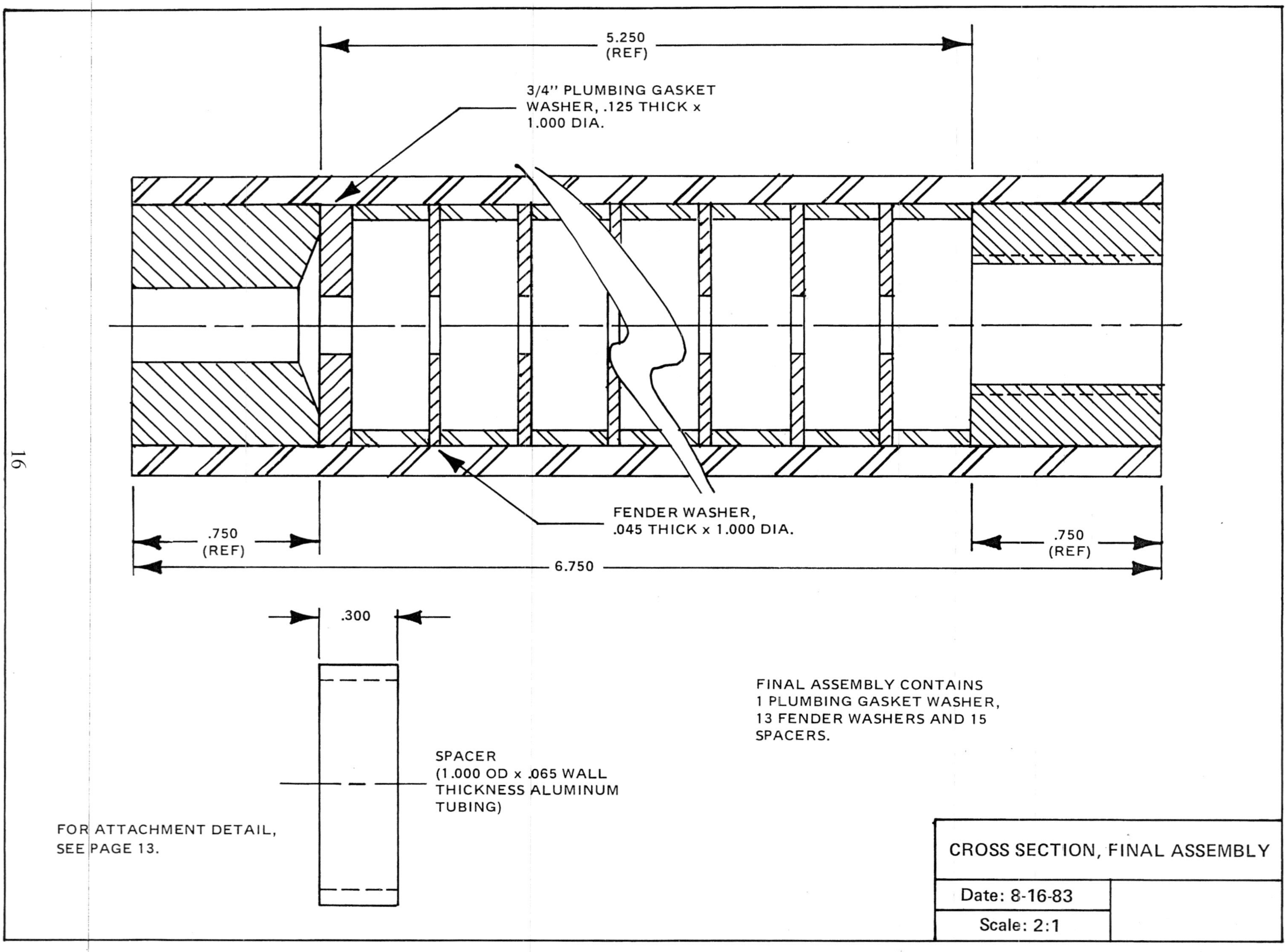
5.250
(REF)
3/4" PLUMBING GASKET
WASHER, .125 THICK x
1.000 DIA.
FENDER WASHER,
.045 THICK x 1.000 DIA.
.750
(REF)
6.750
.750
(REF)
.300
SPACER
(1.000 OD x .065 WALL
THICKNESS ALUMINUM
TUBING)
FOR ATTACHMENT DETAIL,
SEE PAGE 13.
FINAL ASSEMBLY CONTAINS
1 PLUMBING GASKET WASHER,
13 FENDER WASHERS AND 15
SPACERS.
CROSS SECTION, FINAL ASSEMBLY
Date: 8-16-83
Scale: 2:1

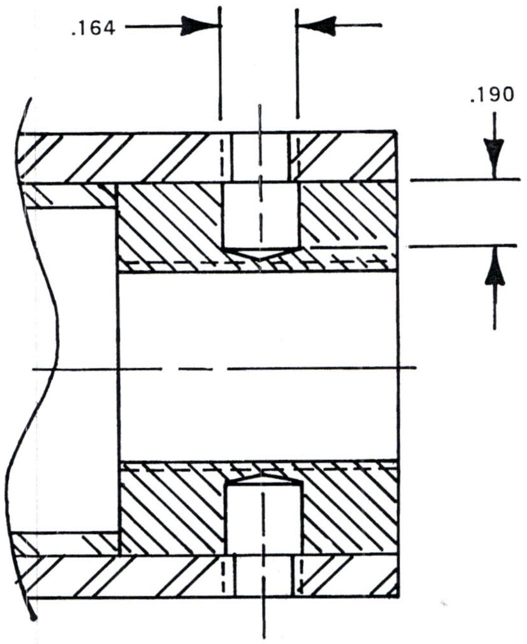

ATTACHMENT DETAIL	
Date: 8-17-83	
Scale: 2:1	

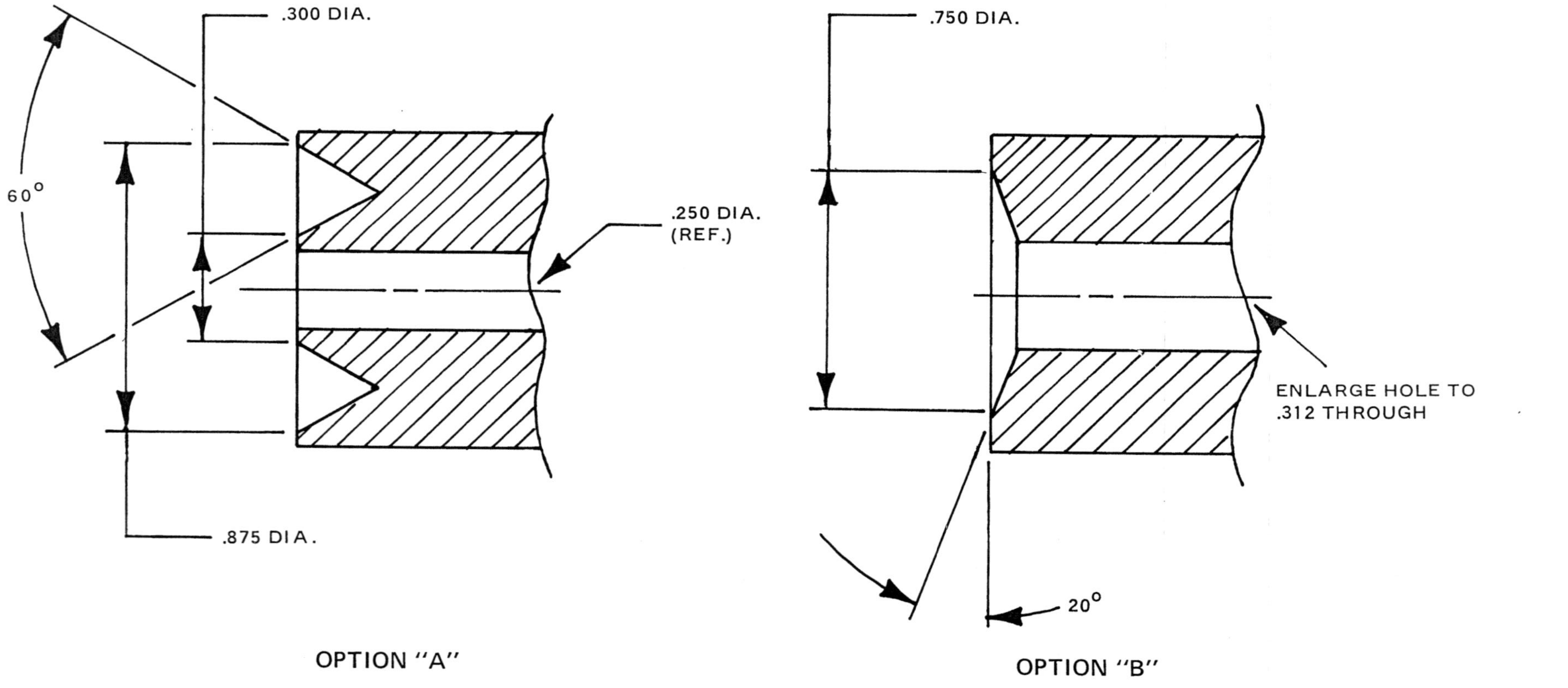

ALTERNATES, FRONT END PLUG

Date: 8-15-83

Scale: 2:1

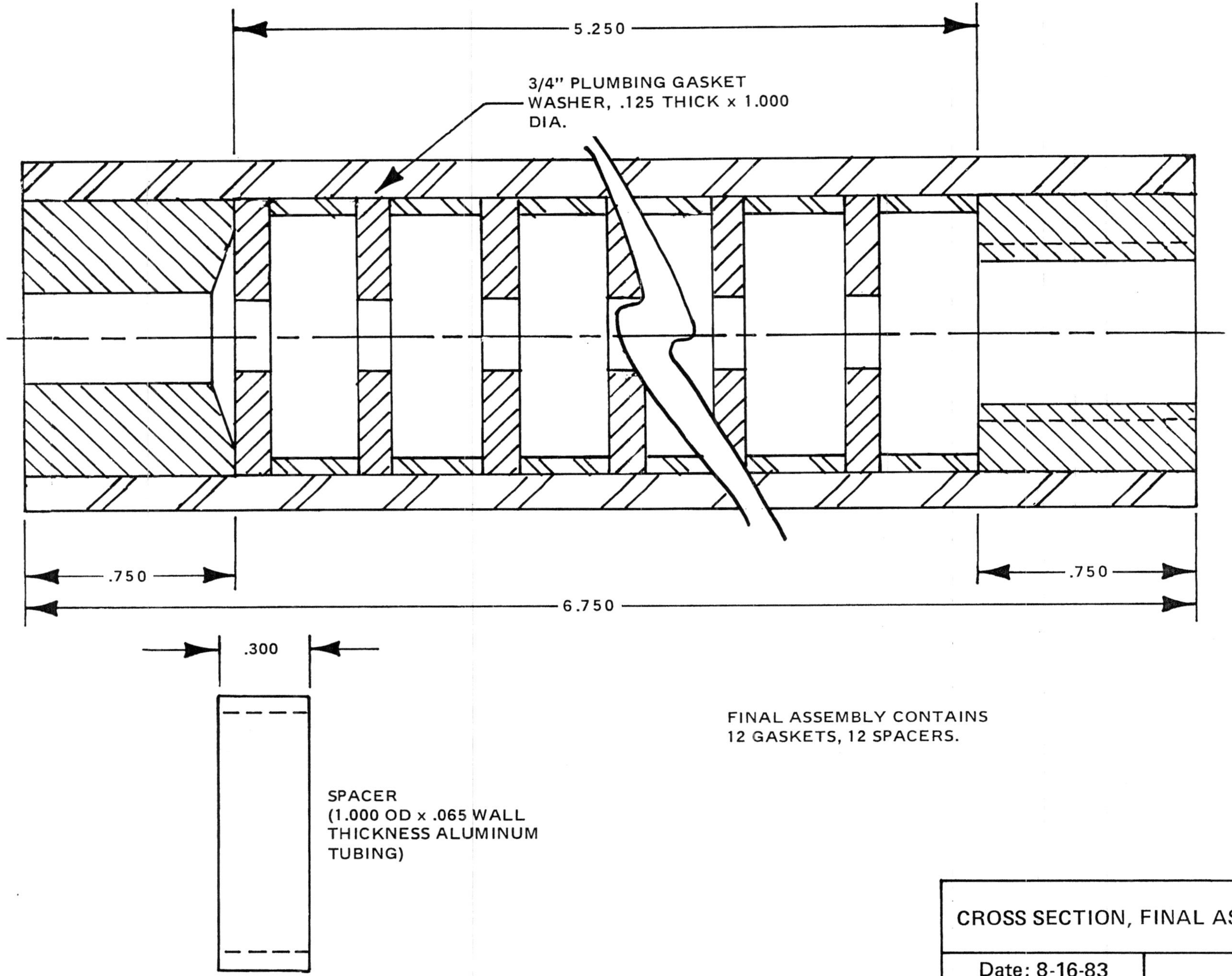

CROSS SECTION, FINAL ASSEMBLY	
Date: 8-16-83	
Scale: 2:1	

A spacer, made from a .300 long piece of 1.000 OD x .065 wall aluminum tubing, separates each fender washer baffle from its neighbor. A 3/4" neoprene plumbing washer which measures 1.000 OD was used for an end wipe (see page 12).

PLUMBING WASHER BAFFLES

Another simple baffle design can be constructed entirely from 3/4" neoprene plumbing washers. Since the bullet must "push" its way through the washers, a "seal" is caused as the bullet passes through each chamber. While this negatively affects accuracy, **it** tends to slow the bullet down to subsonic (1100 fps) speed.

The features of this suppressor design are ease of construction, effectiveness and the ease with which the baffles can be replaced (see page 15).

GAS TRAPPING BAFFLES

This novel design is the most difficult to construct of any contained herein. The baffles are constructed by soldering an eyelet to each side of a fender washer. Using a specially designed soldering fixture, solder a 1/4" grommet eyelet to one side of the washer and a 3/8" eyelet to the other.

Because of the plating on the fender washers, soldering is difficult. To help this situation, the plating can be removed by carefully immersing the washers in a solution of sodium hydroxide (household lye) for a few minutes. Slowly add the lye to water, not the reverse. Wear eye protection and rubber gloves while doing this. Thoroughly wash and dry the fender washers after the plating is removed.

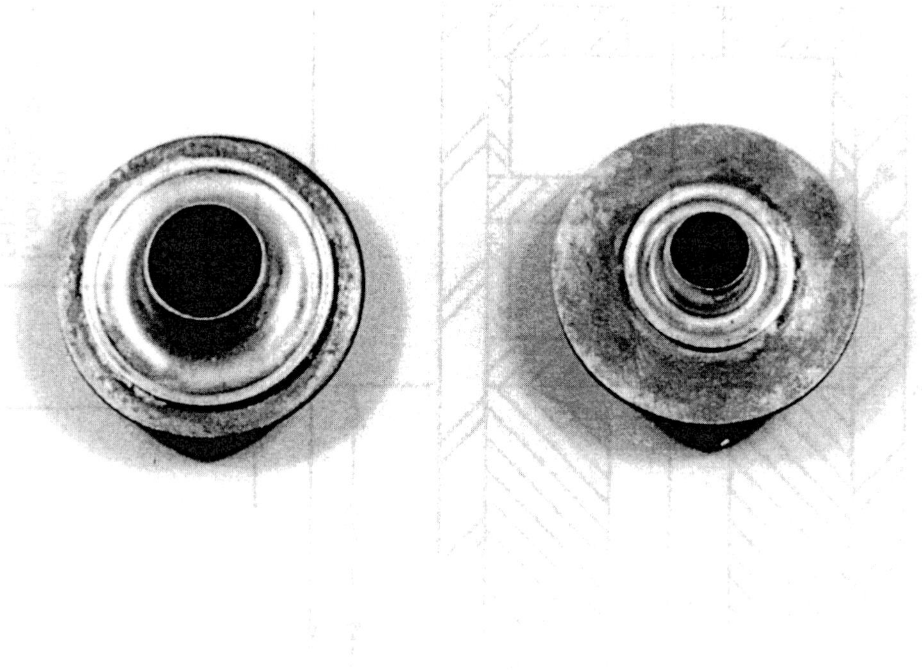

Front and back view of eyelet/baffle assembly after soldering.

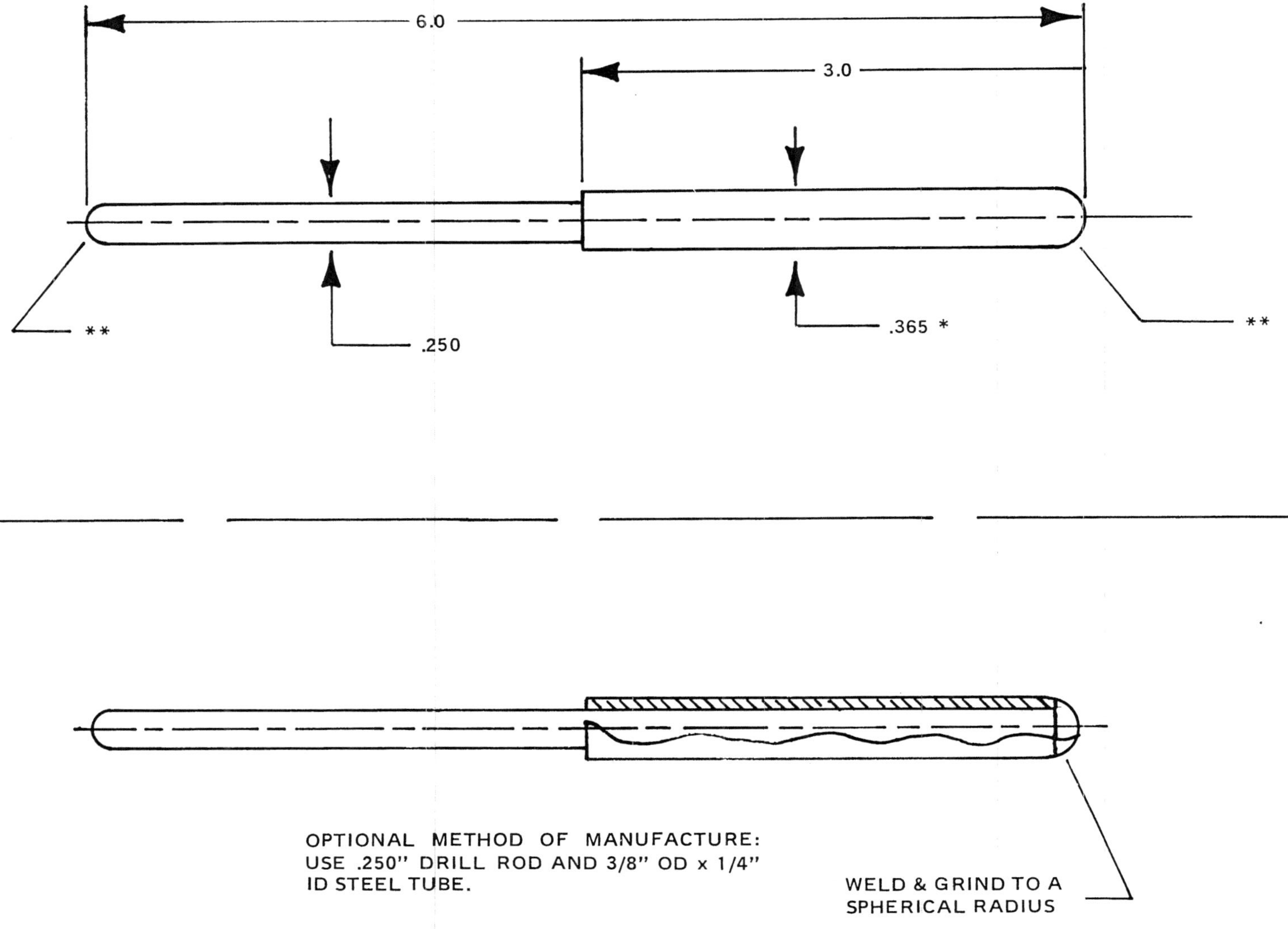

* THIS DIMENSION COULD VARY DEPENDING ON GROMMET ID

** TAPER EACH INTO A SPHERICAL RADIUS

WAFER ASSEMBLY TOOL	
Date: 8-15-83	
Scale: 1:1	

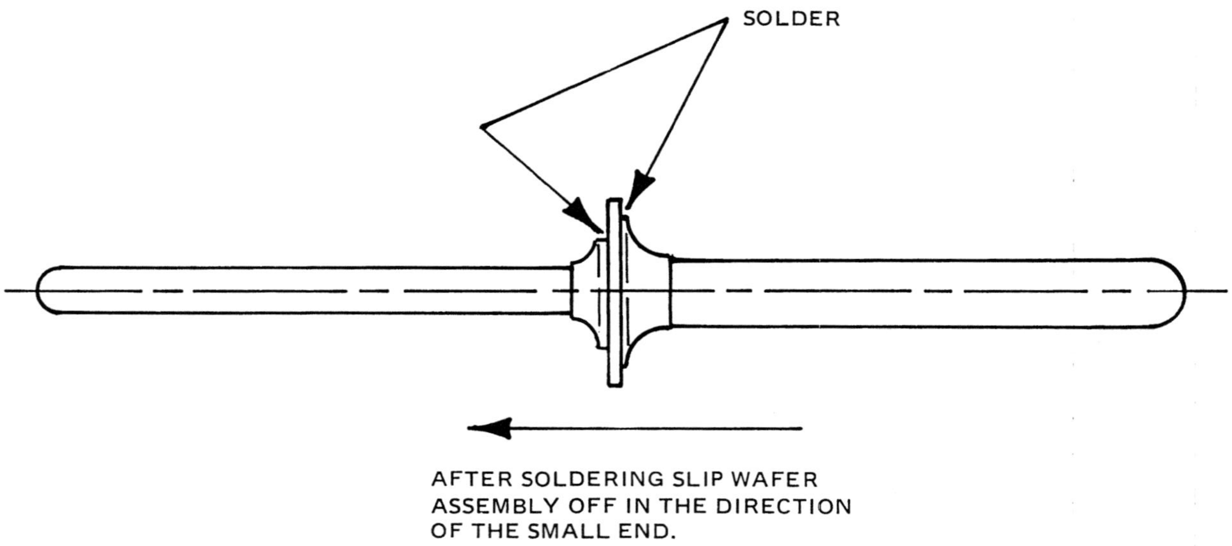

WAFER ASSEMBLY

Date: 8-16-83

Scale: 1:1

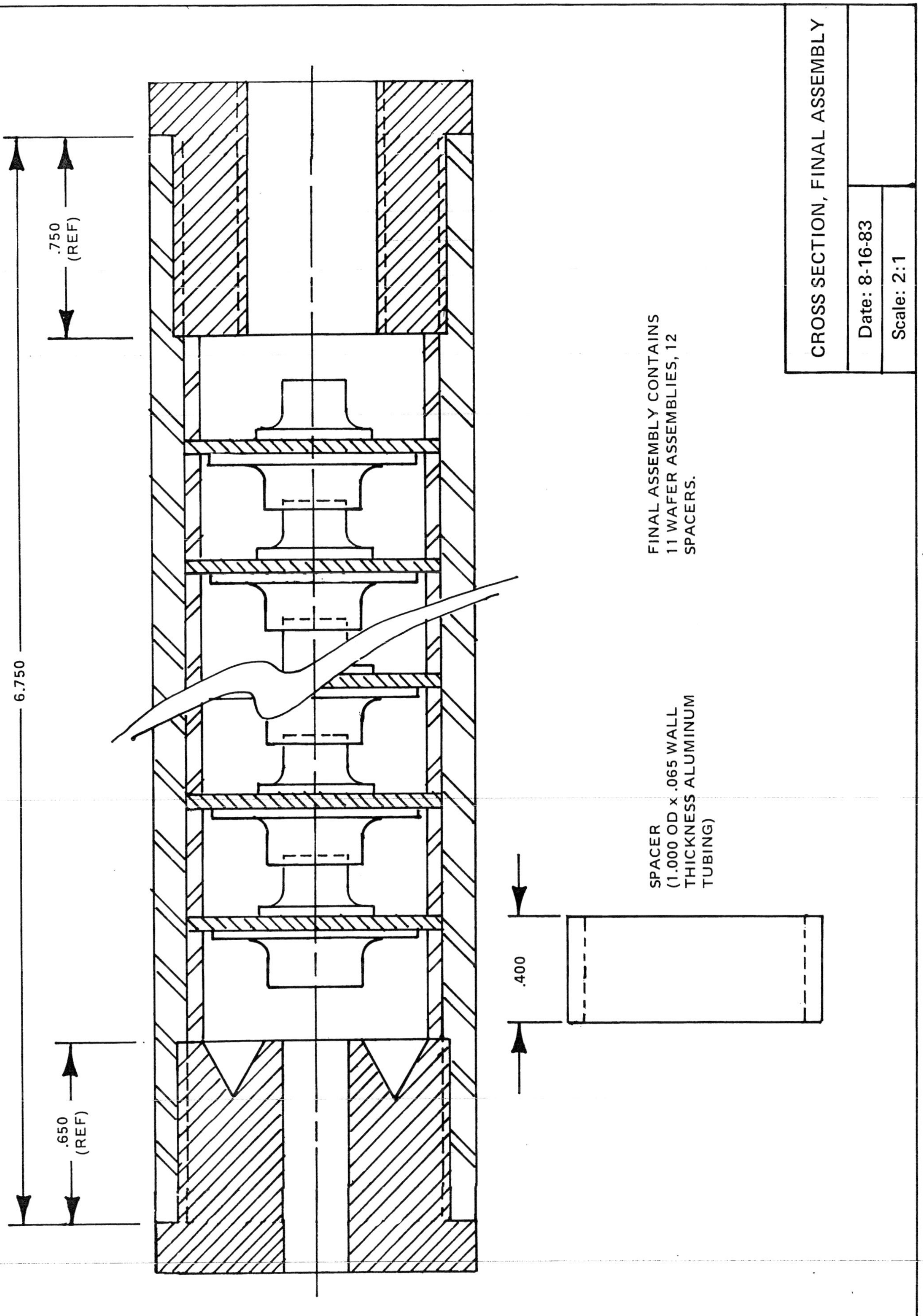
CROSS SECTION, FINAL ASSEMBLY
Date: 8-16-83
Scale: 2:1
FINAL ASSEMBLY CONTAINS
11 WAFER ASSEMBLIES, 12
SPACERS.
SPACER
(1.000 OD x .065 WALL
THICKNESS ALUMINUM
TUBING)
.750
(REF)
6.750
.650
(REF)
.400

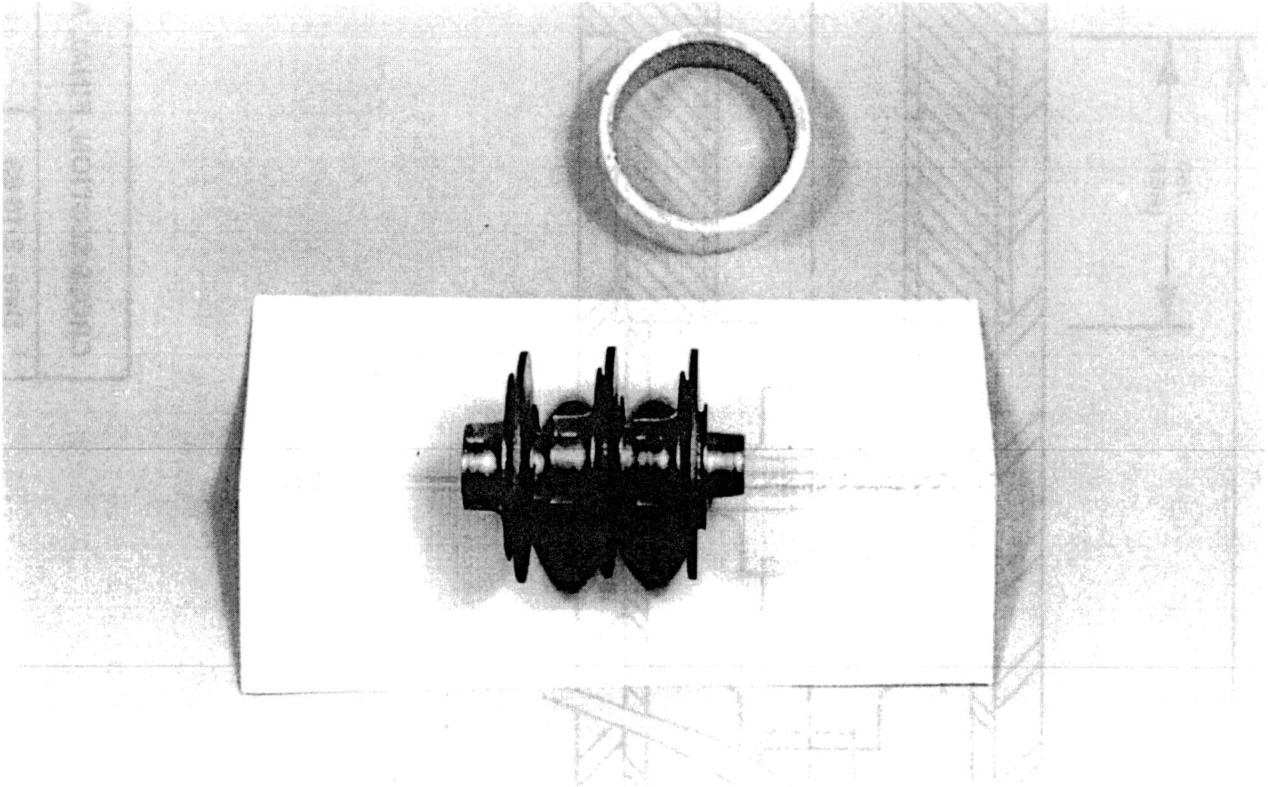

This is how the eyelet/baffle assemblies fit together inside the tube.

Commercially available screen wire suppressor "pack" and the tube for it. Black rubber washers at either end are used merely to hold components onto the wooden dowel.

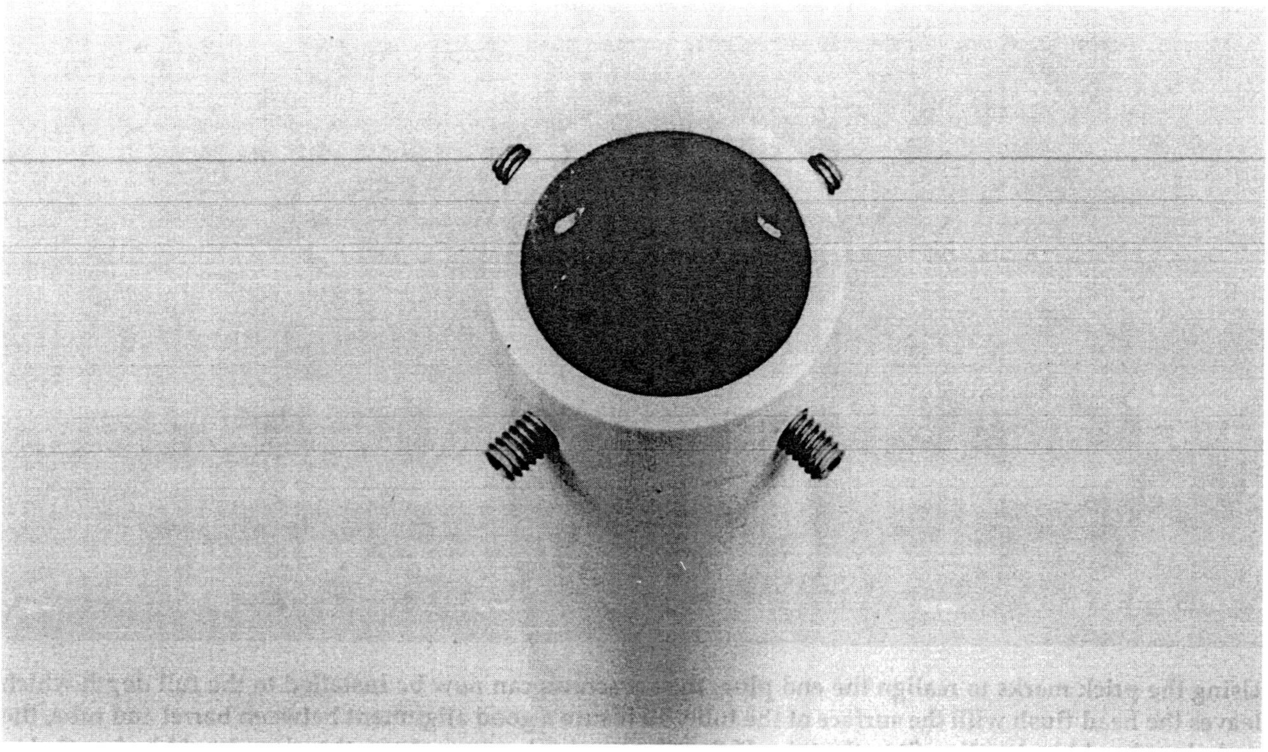

One method for fasting the end plugs to the tube is shown here. Four 8-32 holes are tapped 3/8" from the end of the tube as shown. Install four 5/16" set screws as shown.

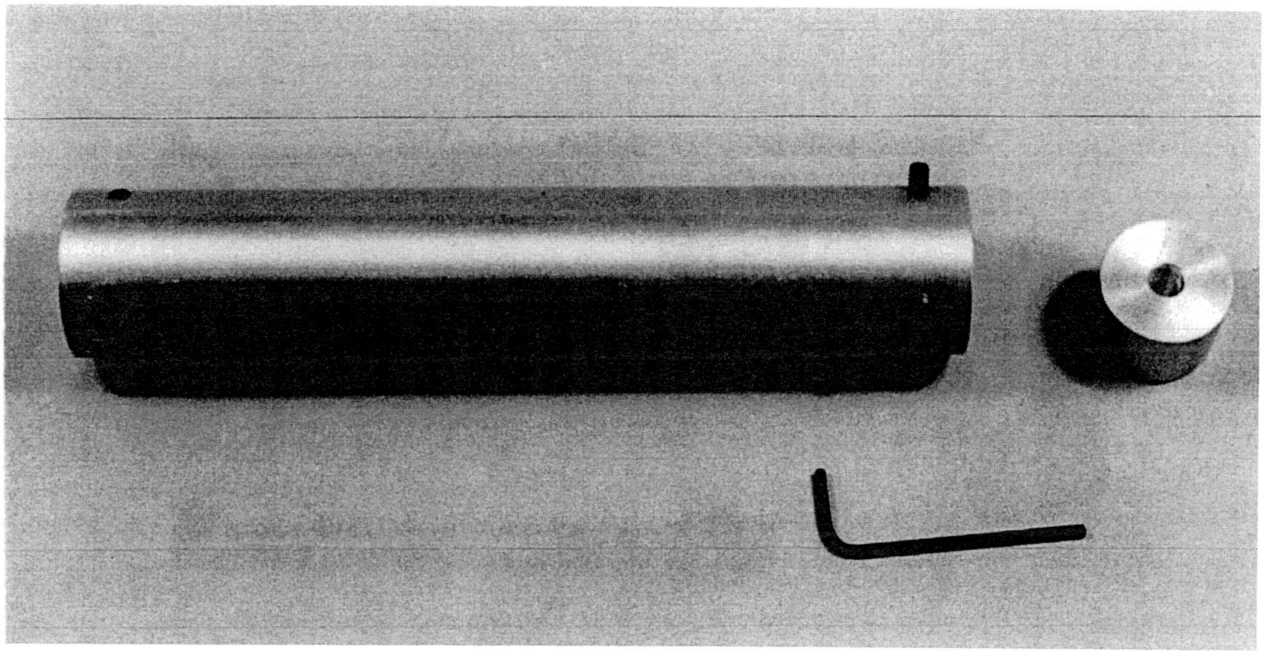

Slip the end plug into the tube until flush and lightly tighten set screws. One at a time, remove the set screws and, using the tapped hole as a guide, drill a shallow starting hole in the plug. Reinstall set screws immediately after drilling. Using a small center punch, make a small prick mark on the end of the tube and plug so it can be oriented when reassembling. Remove the plug and drill the four .161 dia. holes (No. 20) to a depth of 3/16".

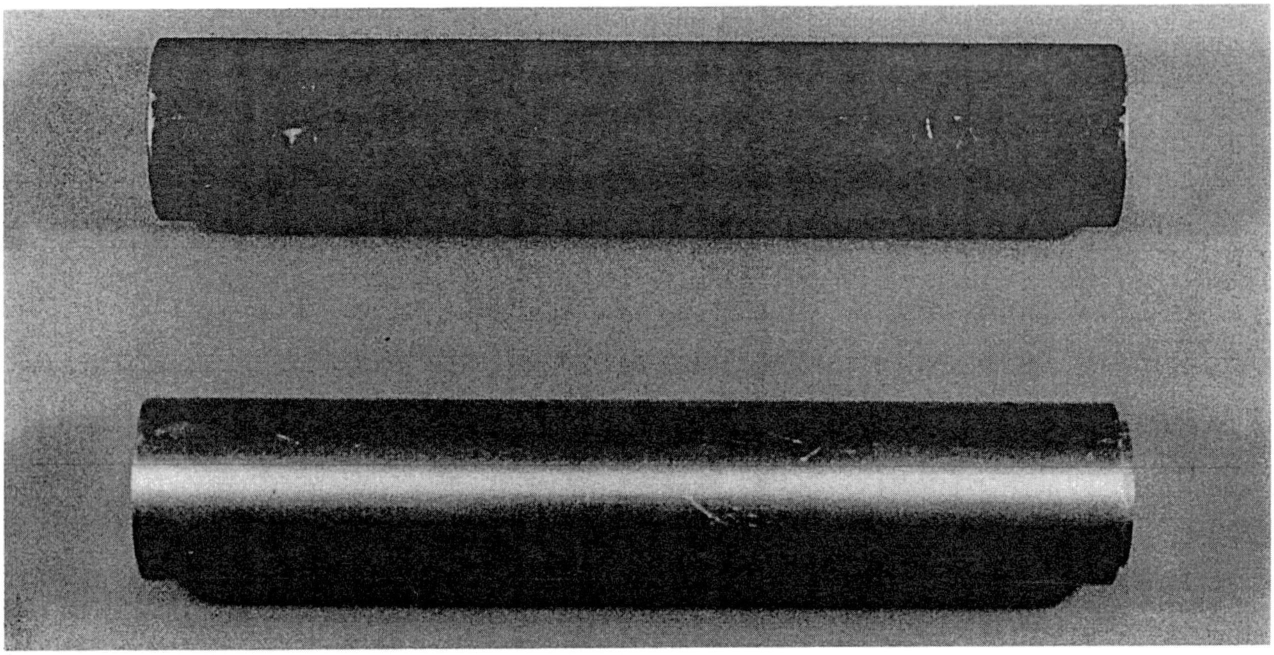

Using the prick marks to realign the end plug, the set screws can now be installed to the full depth which leaves the head flush with the surface of the tube. To insure a good alignment between barrel and tube, the end plug should be in a line fit to the tube. If there's any play between them, the plug should be knurled to increase its diameter.

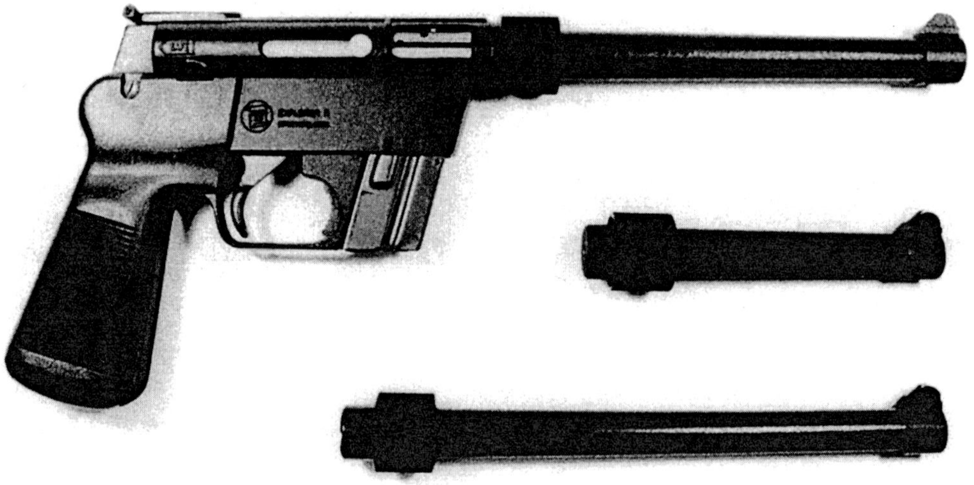

A good, inexpensive pistol to use when experimenting with suppressor design is the Charter Arms Explorer II with its 8" barrel. Extra barrels can be purchased in either 6" or 10" lengths. These barrels are ideal for threading to mount a suppressor.

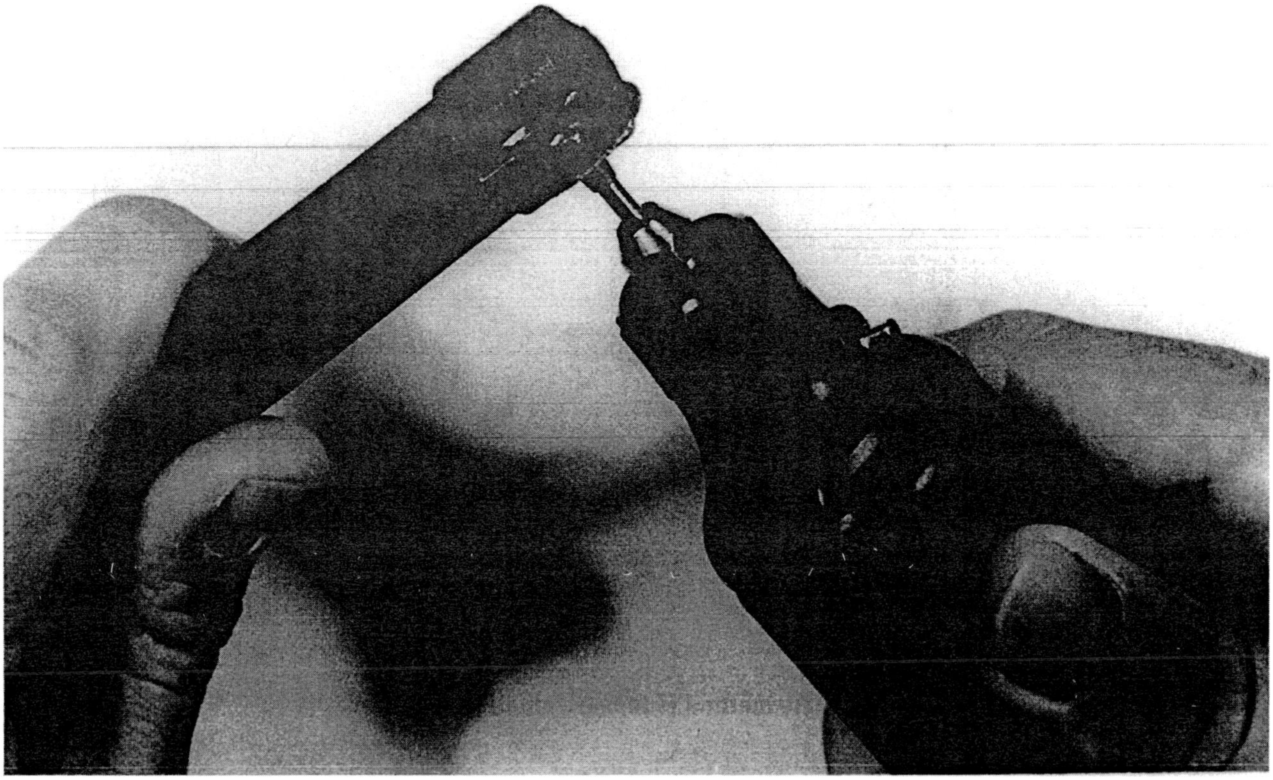

The front sight is pressed over the barrel and can be removed by sawing it almost in two as shown here.

Use a large screwdriver to spread and slip the sight off.

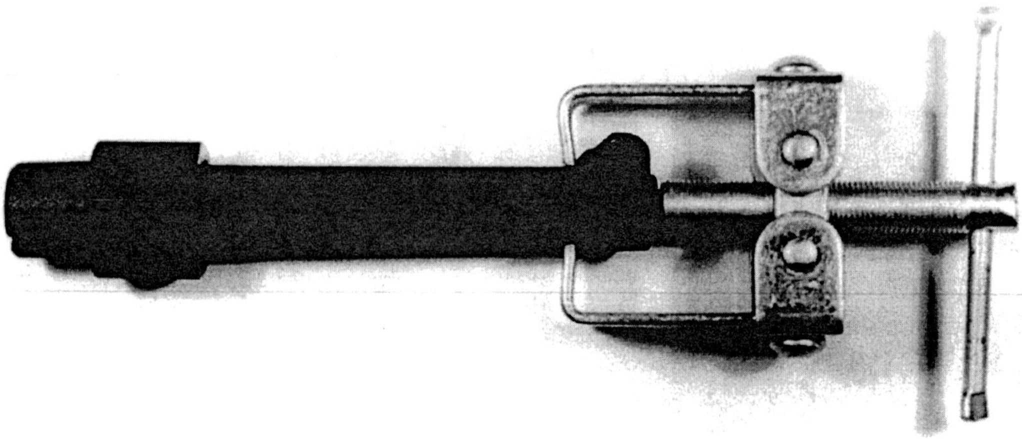

An alternate method is to use a small gear puller.

The soldering fixture is a force fit for both eyelets and this holds them and the fender washer concentric and perpendicular to the axis. Use plumber's soldering paste in areas of contact, and using a small torch, solder them together. After cooling, slip the assembly off the small end of the fixture.

Using a .400 long spacer, these baffles are assembled as shown in the drawing. In use, the bullet enters each chamber by passing through the 1/4" eyelet first. Some of the gas following the bullet is stripped off and momentarily trapped in each chamber.

With the front sight removed, the barrel is pervect to turn down to 1/2" and thread.

Finished barrel and suppressor.

While not shown, the assembly could be altered to allow for an end wipe, again being a 3/4" plumbing washer as described previously.

SCREEN WIRE BAFFLES

While not the easiest to construct, the screen wire baffle suppressor is one of the most efficient. Besides the tube and end plugs, the main component is 300 to 400 screen wire disks 1/4" ID x 1.000 OD. They can be individually cut out by hand, punched out on a simple punch and die set, or purchased as a "pack" from one of the suppliers that advertise in FIREPOWER Magazine. One supplier even sells bulk screen wire for the do-it-yourselfers. Regular brass screen wire could be used. However, somewhat finer mesh is more desirable.

One commercially available screen wire "pack" is illustrated herein. Such packs usually come complete with end plugs. The rear plug is tapped 1/2-28. While this is the thread found on the AR-7s and AR-180, a standard 1/2-20 UNF would be much more desirable. When threading your own end plugs and barrel use 1/2-20 threads.

Note in the photograph on page 20 that the neoprene baffles get progressively closer together toward the front. As we have not tested for optimum spacing, some experimentation can be done on the builder's part, if so desired. However, we have indicated recommended *approximate* dimensions in the drawing on page 27.

FITTING TO THE BARREL

By one means or another, a suppressor can be fitted to just about any .22 rifle or pistol. A weapon that lends itself to ease of suppressor attachment is the Charter Arms Explorer II pistol. Perhaps the ugliest gun in the world, its action is identical to that of the AR-7 rifle. Both the pistol and extra, interchangeable barrels for **it** are inexpensive. By adding an index notch to the AR-7, **it** will accept the pistol barrels (legal with prior BATF approval).

The pistol barrels have the entire front sight pressed on. They can be removed by sawing into the sight band that surrounds the barrel or by using a small gear puller. For best accuracy, the barrel should be threaded in a lathe while being turned between centers.

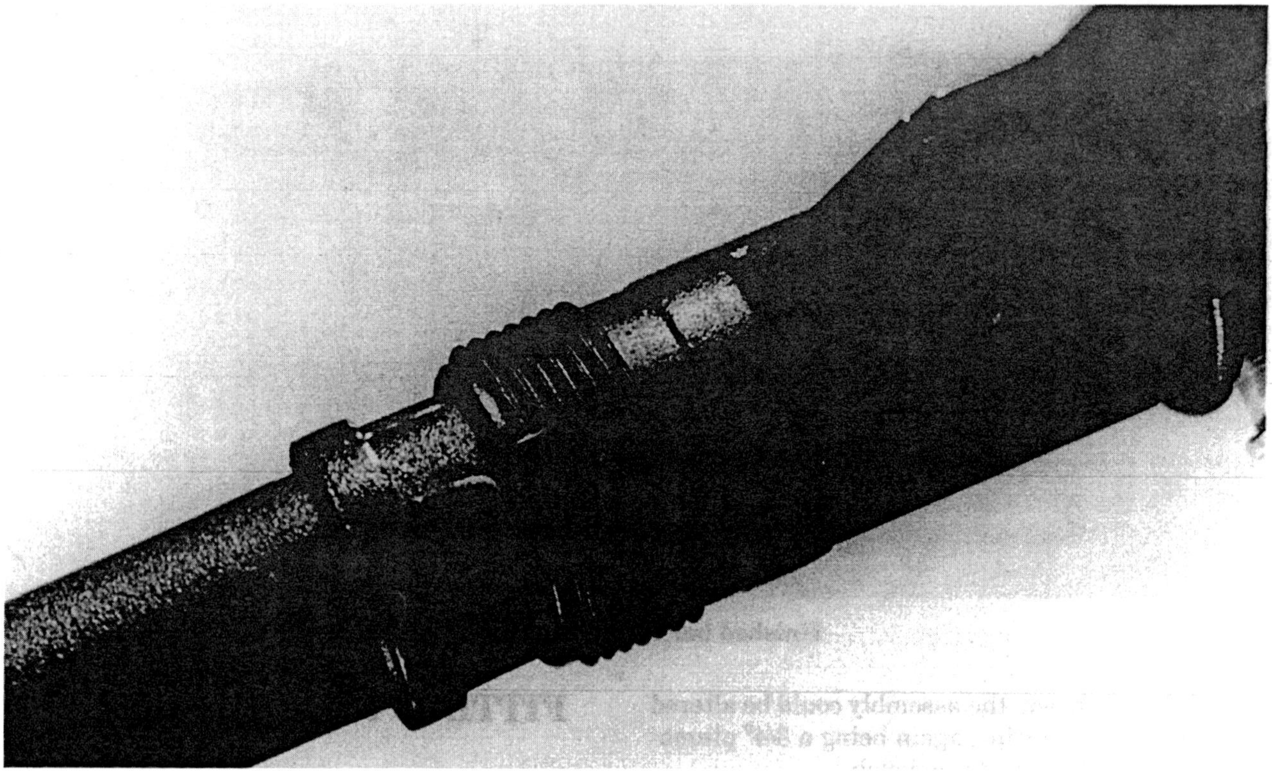

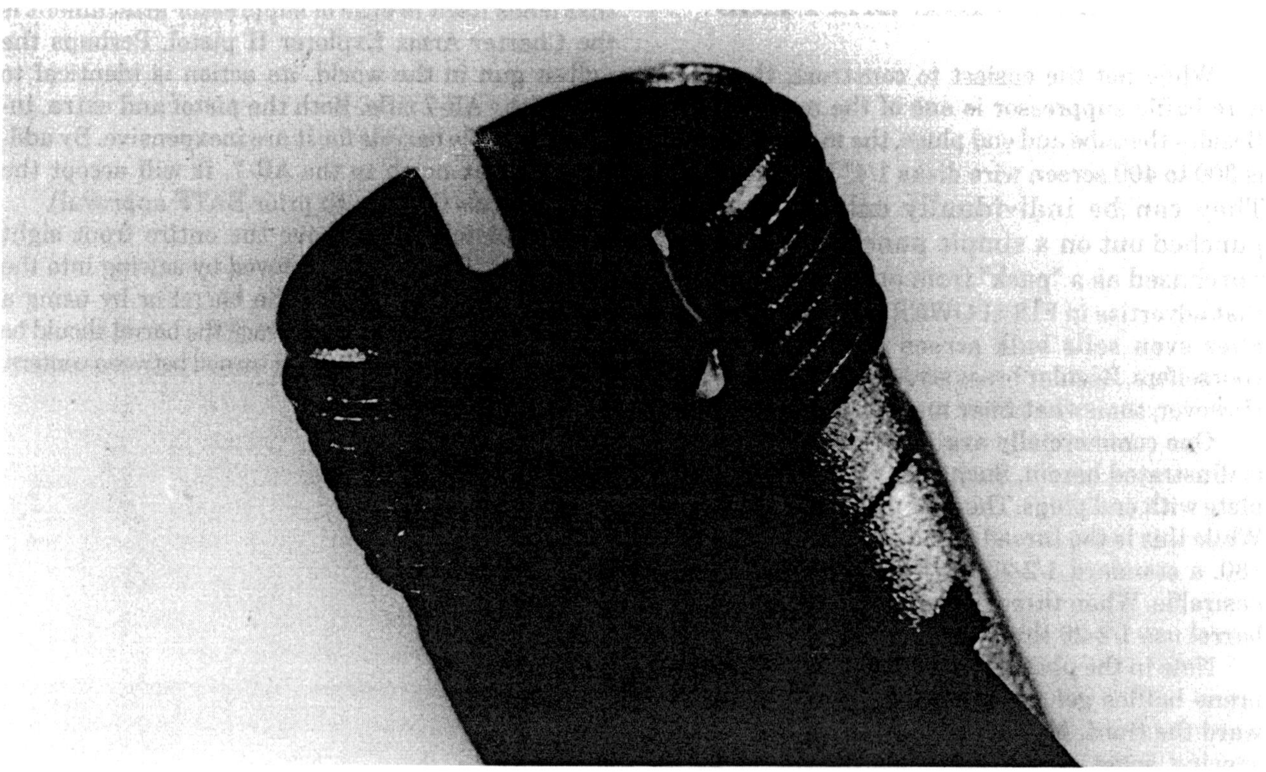

The Explorer II barrels have the indexing "key" located 180° from the indexing slot of the AR-7. By adding another slot directly opposite the existing one the pistol barrels can be installed on the AR-7 rifle.

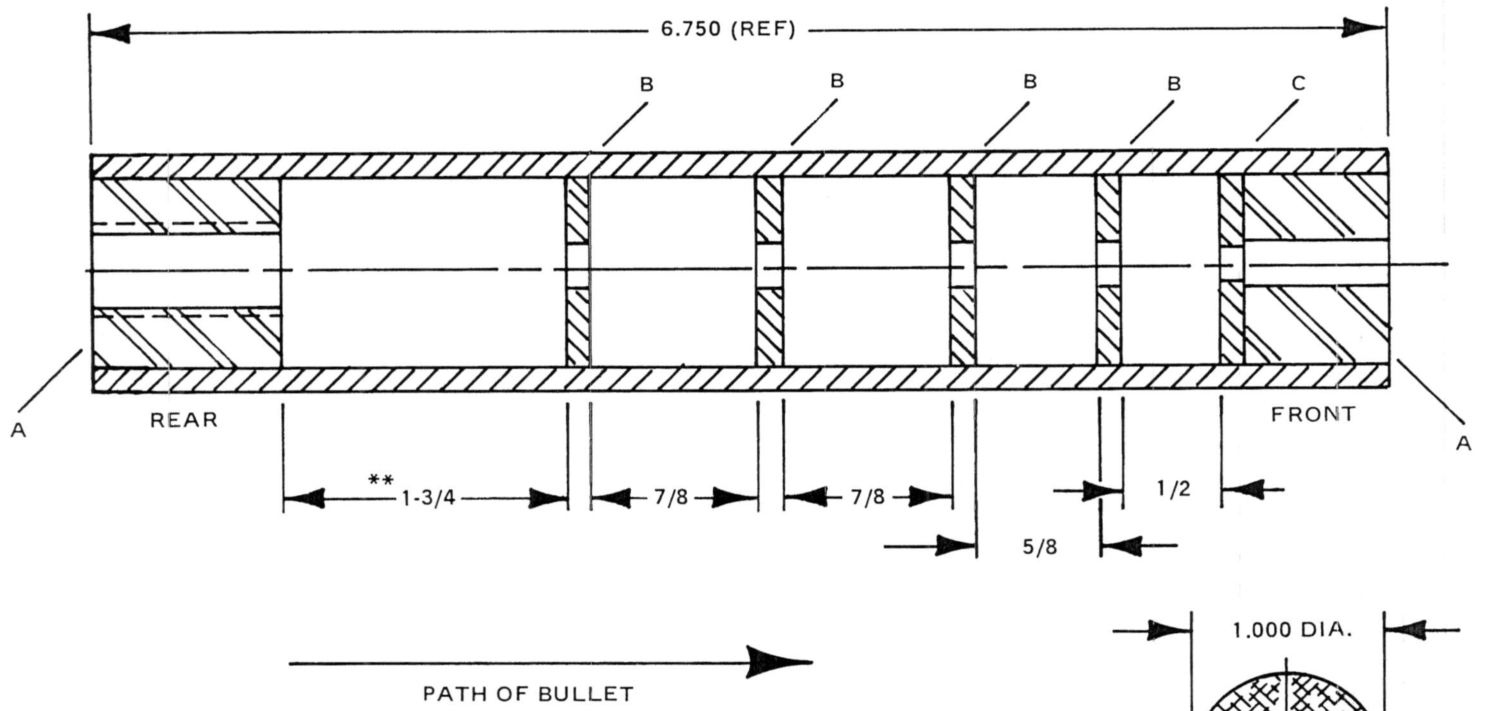

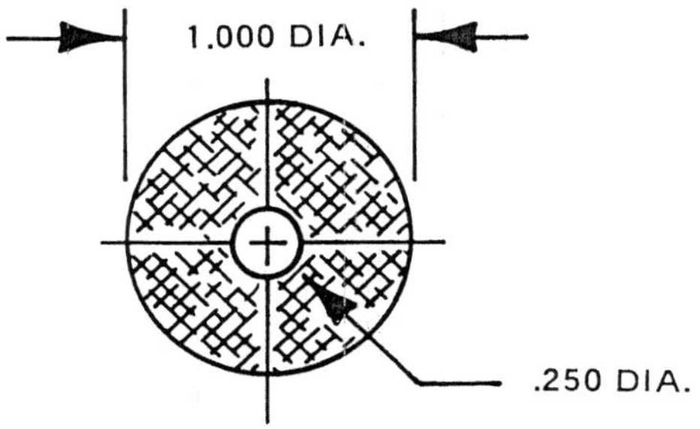

BRASS SCREEN WASHERS
APPROX QTY = 300

A. 3/4" LONG END PLUGS.

B. *3/4" NEOPRENE PLUMBING GASKET WASHER WITH CENTER HOLE ENLARGED TO .250.

C. *3/4" NEOPRENE PLUMBING GASKET WASHER.

* 3/4" GASKETS MEASURE 3/16" ID x 1" OD x 1/8" THICK.

** APPROX DIMS

CROSS SECTION, FINAL ASSEMBLY	
Date: 8-16-83	
Scale: 1:1	

Charter Arms AR-7 Explorer fitted with Explorer II pistol barrel and suppressor. The suppressor shown in the photo can utilize any of the three homemade baffle types outlined in the text.